# What Is the World Made Of?

## All About Solids, Liquids, and Gases

by Kathleen Weidner Zoehfeld • illustrated by Paul Meisel

HarperCollins*Publishers*

*Special thanks to Dr. Leonard Fine of Columbia University for his expert advice*

The art in this book was created using a mixed-media technique that includes pen and ink,
watercolor, acrylic colored pencil, and pastel on Arches hot press paper.

The *Let's-Read-and-Find-Out Science* book series was originated by Dr. Franklyn M. Branley, Astronomer Emeritus and former Chairman of the American Museum–Hayden Planetarium, and was formerly co-edited by him and Dr. Roma Gans, Professor Emeritus of Childhood Education, Teachers College, Columbia University. Text and illustrations for each of the books in the series are checked for accuracy by an expert in the relevant field. For more information about Let's-Read-and-Find-Out Science books, write to HarperCollins Children's Books, 10 East 53rd Street, New York, NY 10022, or visit our web site at http://www.harperchildrens.com.

What Is the World Made Of?
All About Solids, Liquids, and Gases
Text copyright © 1998 by Kathleen Weidner Zoehfeld
Illustrations copyright © 1998 by Paul Meisel

Library of Congress Cataloging-in-Publication Data
Zoehfeld, Kathleen Weidner.
   What is the world made of? : all about solids, liquids, and gases / by Kathleen Weidner Zoehfeld ; illustrated by Paul Meisel.
      p.      cm. — (Let's-read-and-find-out science.    Stage 2)
   Summary: In simple text, presents the three states of matter, solid, liquid, and gas, and describes their attributes.
   ISBN 0-06-027143-4. — ISBN 0-06-027144-2 (lib. bdg.). — ISBN 0-06-445163-1 (pbk.)
   1. Matter—Constitution—Juvenile literature.   [1. Matter.]   I. Meisel, Paul, ill.   II. Title.   III. Series.
QC173.16.Z64   1998                                                                                           97-30658
530.4—dc21                                                                                                        CIP
                                                                                                                   AC

Typography by Elynn Cohen and Christine Casarsa
1   2   3   4   5   6   7   8   9   10

First Edition

# What Is the World Made Of?

## All About Solids, Liquids, and Gases

Have you ever seen anyone walk through a wall?

4

Did you ever drink a glass of blocks?

Have you ever played with a lemonade doll, or put on milk for socks?

THANKS!

5

Walls and blocks, dolls and socks. Milk and lemonade.
Rocks and trees. All of these things are made of matter.

The air in the breeze that blows the leaves. Water flowing in the creek. Everything on earth is made of matter. Lucky for us, it's easy to tell that not all matter is alike.

Matter comes in three states. It can be solid, liquid, or gas.

Walls and blocks and socks are SOLIDS.

Milk and lemonade and water are LIQUIDS.

AIR

LUNGS

The air you breathe is a GAS.

Some solids are hard and some are soft. But all solids hold their shape unless you do something to change them.

A wooden block is hard. You can push it and pull it and squeeze it—it will always stay squared. If your baby brother pounds it with his toy truck, it may chip or break. But then the broken pieces will hold their shapes.

9

Modeling clay is soft. You can roll it out flat, like a pancake, or squeeze it into a ball. You can cut it with a knife or scissors. But if you leave it alone, it will hold whatever shape it is in. It is a solid.

Liquids have no shape. You pour a glass of milk for your little sister, and the milk takes on the shape of the glass. If she knocks it over, the milk spreads out on the tabletop. It flows over the edge like a waterfall. It drips and spatters on the floor. Milk is not round or square—it has no shape at all.

Liquids can be thick, like a milkshake, or thin, like water. They can feel slippery, like cooking oil, or sticky, like maple syrup. But all liquids can be poured. And all liquids take on the shape of whatever they are in.

Gases have no shape either. Like air, most gases are invisible—you cannot see them. But you can feel them. Hold out your arms and spin. You can feel the air move against your skin. Air fills up all the space around you.

15

Gases spread out to fill up any container they are in, no matter how big. Ask a grown-up if you can borrow a bottle of perfume. Then ask someone to be your assistant.

Go into a room and close all the doors and windows. Stand in one corner and have your assistant stand in the opposite corner. Open the bottle and wave it around gently.

Can your friend smell the perfume? How long does it take before she can smell it?

When you open a bottle of perfume, some of the gas in the perfume escapes. In a few minutes the gas will spread out to every corner of the room.

18

Water is a special type of matter that can change easily from a liquid to a gas. It can also change easily from a liquid to a solid and back again. When matter changes form, we say it is changing its state. A change in temperature is what usually causes matter to change its state.

Whenever you make ice cubes, you are changing water from a liquid to a solid state. To make ice cubes, all you have to do is pour water into an ice-cube tray and put the tray in the freezer. Then wait.

The cold air in the freezer cools the water. In a few hours the liquid water gets so cold that it freezes. It is solid.

Pop an ice cube out of the tray. How does it feel? Push it and pull it and squeeze it—it is as solid as one of your wooden blocks. If you hold it in your hand for a while, though, you will feel the ice turning into water again.

The warmth from your hand makes the ice melt. Liquid water begins to drip between your fingers and run up your sleeve.

Hey! Mine is Melting!!

22

With a grown-up's help, you can watch water turning from a liquid to a gas state. Fill a small saucepan about halfway with water. Put the saucepan on the stove, and let the grown-up turn on the burner. After a while, heat from the burner will make the water hot.

Look carefully, and you will see small bubbles forming in the bottom of the pan. As it is heated, some of the water turns into bubbles of gas. The gas is called water vapor.

Soon the bubbles begin to rise and pop. The water is boiling. As the bubbles pop, the water vapor moves up into the room. You might see wisps of whitish steam rising above the water. The steam is concentrated water vapor. (Do not try to touch the steam: It is very hot and could burn your skin.)

Like all gases, the water vapor gradually spreads throughout the room you are in. If the door is open, it will keep on spreading out the door. You cannot see it, but the water vapor is there all around you.

If you want to see it again, you have to change it back into water. Fill a drinking glass with water and ice cubes. (This works better if you break the cubes into small pieces.) Make sure the outside of the glass is dry. Before long you will see droplets of water forming on the outside of the glass.

The ice makes the glass cold. The cold glass cools the water vapor that is in the air next to it. As the water vapor cools, it turns back into liquid water again. It gathers in drops on the outside of the glass.

All matter, everything on earth, is either solid, liquid, or gas. Water changes its state easily as it gets warmer or colder. But most things stay in one state or another. Solids stay solid. Liquids stay liquid. Gases stay gaseous.

30

And it's a good thing they do! Can you imagine a world where your toys melt when it gets too hot? Where the walls of your house turn into hazy gas, and animals just walk in and out as they please? A place where, on cold days, you have to swim through the air, and where everything you'd like to drink is hard as a block? What a crazy world it would be!

OUCH!

# FIND OUT MORE ABOUT MATTER

- **D**oes all matter take up space? You know that all solids take up space. To find out about liquids, all you have to do is take a bath.

  1. Stick a piece of tape to the inside of the tub and draw a pencil line on it.

  2. Fill the tub so that the water is even with the pencil line. Climb into the tub. The water will rise above the line. It takes up space, and so do you. You and the water cannot be in the same space at the same time.

- **D**o gases take up space? Find out with a large bowl, a small drinking glass, a paper napkin, and some water.

  1. Fill the bowl with water until it is about three quarters full.

  2. Crumple the napkin and push it down into the bottom of the glass.

  3. Turn the glass upside down and push it straight down into the bowl. The napkin doesn't get wet because the glass is full of air. The air and the water cannot be in the same space at the same time.

- **M**ake a chart that matters.

  1. Take a large sheet of paper and fold it in half and then in half again. Unfold the paper. You've got four sections. Four? But there are only three states of matter! Don't worry. Label one section SOLIDS, one LIQUIDS, and one GASES. Leave the fourth section blank for now. Then choose a day for looking.

  2. Wherever you go, notice everything you can. Decide if the object is a solid, liquid, or gas. Then write its name down, or draw a picture of it, in the right section.

  3. If you're looking very carefully, you'll notice that some things are mixtures of solid and liquid, or liquid and gas. Noodle soup, for example, is a mixture of broth (liquid), noodles (solids), and the steam rising above it (gas). So label your fourth section MIXTURES and keep looking!